Field Trips

At the Factory

By Sophie Geister-Jones

level 2 little blue readers

www.littlebluehousebooks.com

Little Blue House is distributed by North Star Editions:
sales@northstareditions.com | 888-417-0195

Produced for Little Blue House by Red Line Editorial.

Photographs ©: SeventyFour/iStockphoto, cover; RicAguiar/iStockphoto, 4, 19; DragonImages/iStockphoto, 7; Morsa Images/iStockphoto, 8–9; Chalffy/iStockphoto, 10; pierrephoto/iStockphoto, 13; Yumi mini/iStockphoto, 14–15; Traimak_Ivan/iStockphoto, 16; 1001nights/iStockphoto, 20–21; Tramino/iStockphoto, 23 (top); jetcityimage/iStockphoto, 23 (bottom); Jr images/Shutterstock Images, 24 (top left); BlackJack3D/iStockphoto, 24 (top right); kvsan/Shutterstock Images, 24 (bottom left); Minerva Studio/Shutterstock Images, 24 (bottom right)

Library of Congress Control Number: 2019908253

ISBN
978-1-64619-027-0 (hardcover)
978-1-64619-066-9 (paperback)
978-1-64619-105-5 (ebook pdf)
978-1-64619-144-4 (hosted ebook)

Printed in the United States of America
Mankato, MN
012020

About the Author

Sophie Geister-Jones likes reading, spending time with her family, and eating cheese. She lives in Minnesota.

Table of Contents

At the Factory

We are on a field trip to a factory.
We want to learn how cars are made.

We have to wear hard hats in the factory. Hard hats keep our heads safe.

hard hat

Many people work at the factory.
They do many different jobs.
They work together to make cars.

Car Parts

Workers use many kinds of tools.
Some of the tools hang from the ceiling.

Cars are made of many parts. We see the frame. The frame gives the car its shape.

frame

We see little parts too.
A battery gives the
car power.
We see a worker test
the battery.

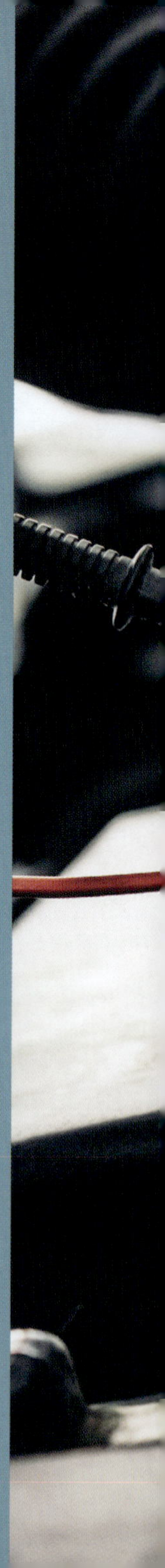

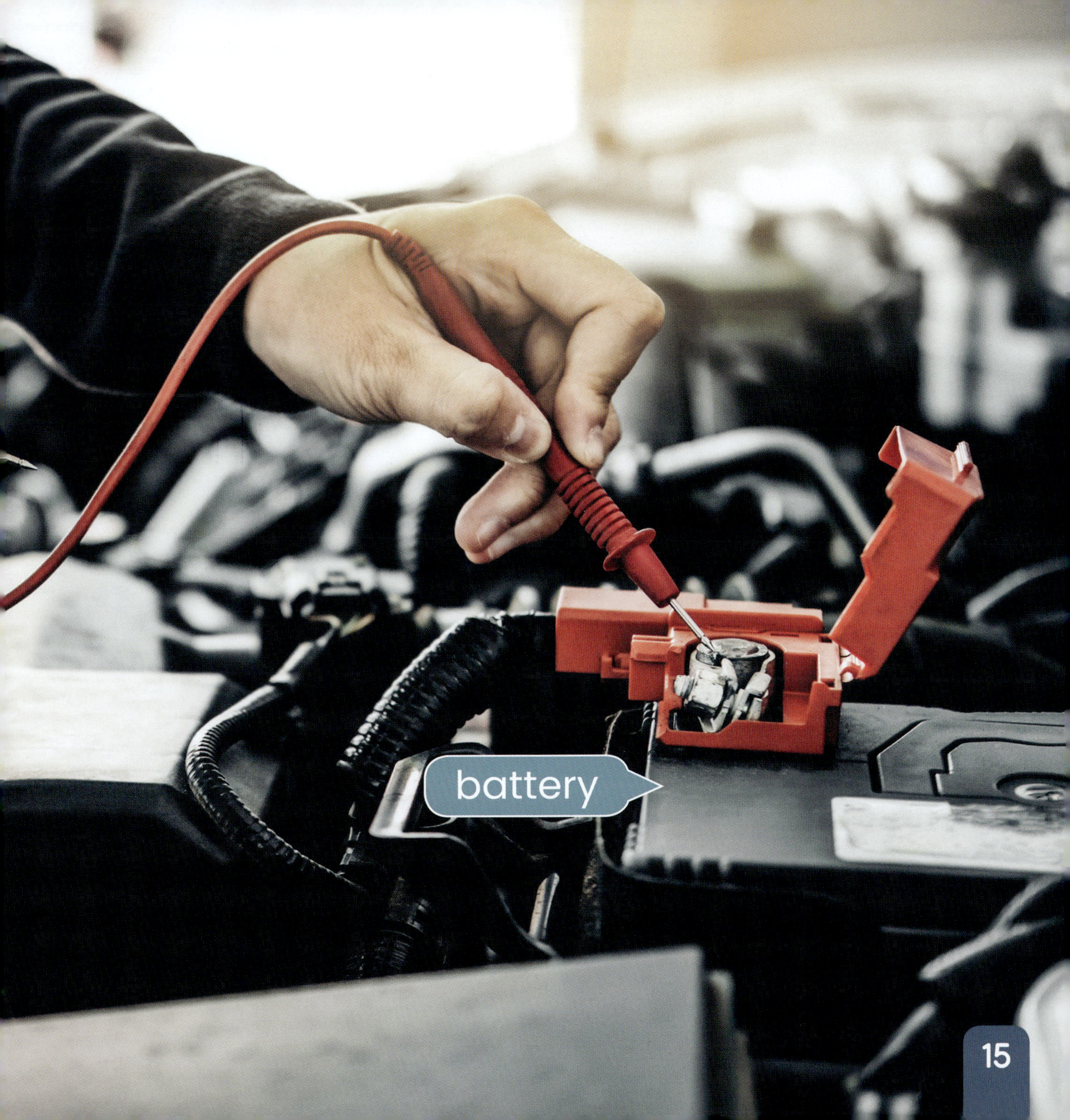
battery

Putting It Together

The cars move in a line. New parts are added to the cars as they move.

The factory has many robot arms.
The robots put some of the parts together.

robot arm

Workers put wheels on the car.

They add the windows.

They make sure everything is in place.

wheel

The cars are sent away on a truck.
They will be sold all over the world.

W&LINA
SVENSTRUP &
VENDELBOE 4/9
AN 59 340
Kässbohrer
1AP 0922

Center

Glossary

battery

wheels

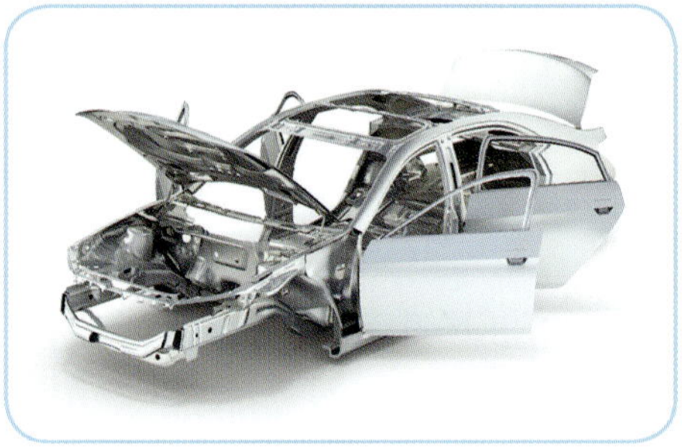

frame

worker

Index